**Crawly**

# SCORPIONS

## LYN SIROTA

**BLACK
RABBIT
BOOKS**

BOL

Bolt is published by Black Rabbit Books
P.O. Box 3263, Mankato, Minnesota, 56002.
www.blackrabbitbooks.com
Copyright © 2020 Black Rabbit Books

Marysa Storm, editor; Grant Gould, designer;
Omay Ayres, photo researcher

Names: Sirota, Lyn A., 1963- author.
Title: Scorpions / by Lyn Sirota.
Description: Mankato, Minnesota : Black Rabbit Books, [2020] | Series:
Bolt. Crawly creatures | Audience: Age 9-12. | Audience: Grade 4 to 6. |
Includes bibliographical references and index.
Identifiers: LCCN 2018020663 (print) | LCCN 2018021643 (ebook) |
ISBN 9781680728187 (e-book) | ISBN 9781680728125 (library binding) |
ISBN 9781644660232 (paperback)
Subjects: LCSH: Scorpions–Juvenile literature.
Classification: LCC QL458.7 (ebook) | LCC QL458.7 .S57 2020 (print) |
DDC 595.4/6–dc23
LC record available at https://lccn.loc.gov/2018020663

Printed in the United States. 1/19

# CONTENTS

# CHAPTER 1

# Meet the SCORPION

A scorpion cools off in the shade. Something nearby moves. The scorpion's hairs sense the **vibrations**. The creature moves again. The scorpion's many eyes spy the movement. It's a wiggly worm. The scorpion strikes. It grasps the worm with pincers. A quick sting and the worm is **paralyzed**. Time for dinner!

## COMPARING LENGTHS

emperor scorpion

fattail scorpion

deathstalker scorpion

Middle Eastern scorpion

inches                    1

6

# A Crawly Creature

There are thousands of kinds of scorpions. These **arachnids** are powerful **predators**. They have eight legs. Their bodies can be yellow, brown, or black. Scorpions blend in with their surroundings. Hard **exoskeletons** keep them safe from attackers.

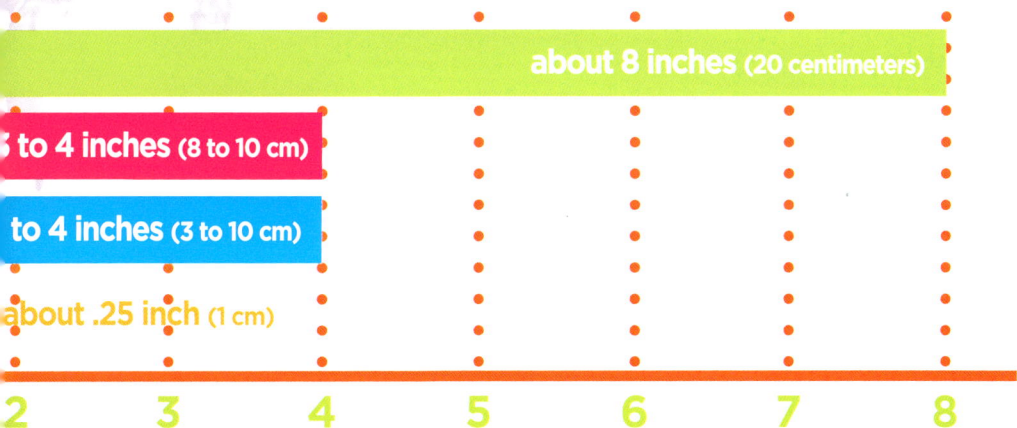

about 8 inches (20 centimeters)

to 4 inches (8 to 10 cm)

to 4 inches (3 to 10 cm)

about .25 inch (1 cm)

2    3    4    5    6    7    8

**STINGER**

**PINCERS**

**EYES**

**ABDOMEN**

**LEGS**

# Scorpion Venom

Each kind of scorpion makes its own **venom**. Some scorpion venom is strong enough to kill **prey**. Other venom only paralyzes. Most kinds aren't deadly to people.

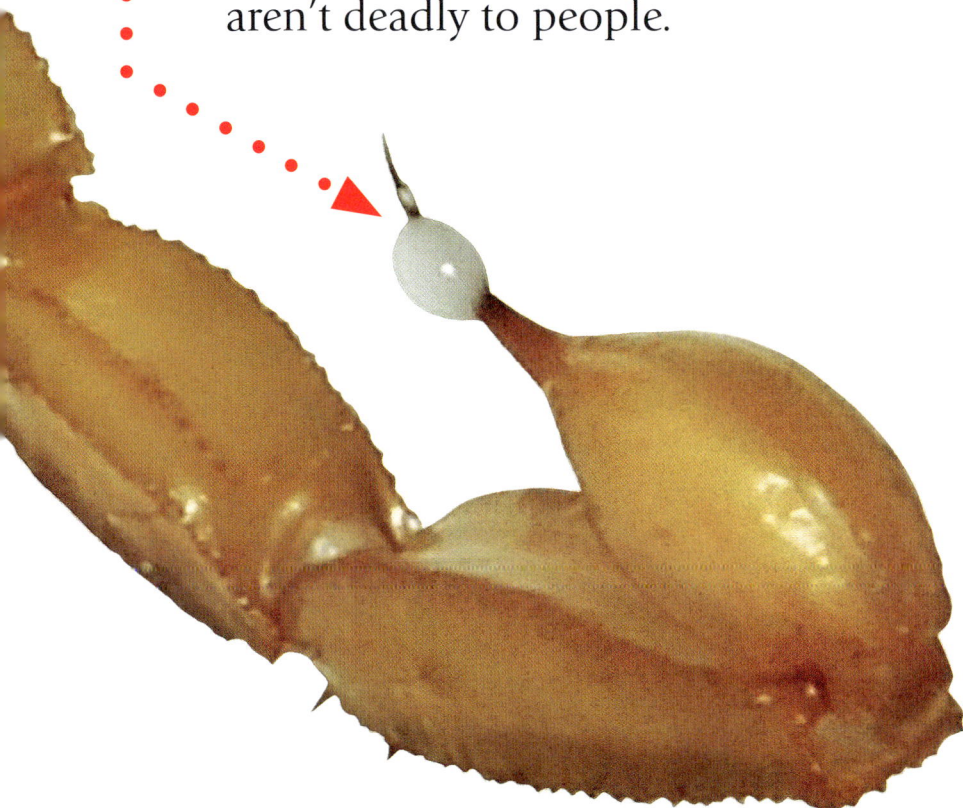

Scientists found frozen scorpions. But they weren't dead! After **thawing** out, they walked away.

# CHAPTER 2

# WHERE THEY LIVE
## and What They Eat

Scorpions are found around the world. They live in deserts, forests, and mountains. Many make their homes under rocks. Extreme temperatures? No problem! In hot deserts, they hide beneath the sand. Staying underground helps them stay cool.

# SCORPION RANGE MAP

Scorpions live on every continent except Antarctica.

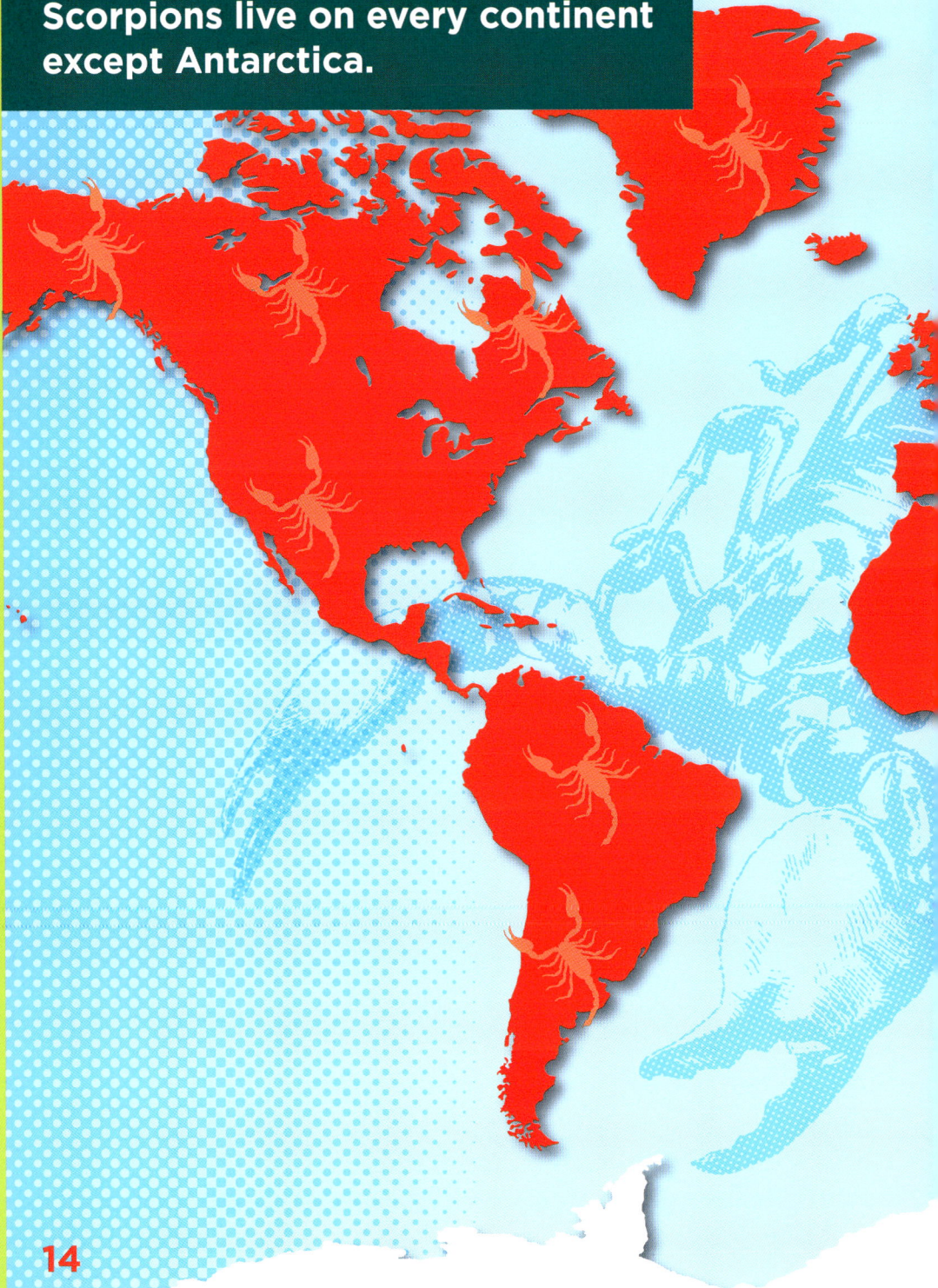

# Nighttime Hunters

Scorpions are most active at night. In darkness, they attack insects and spiders. Many scorpions wait until prey comes near. Then they strike.

Scorpions take their time eating. They must break down their food. Using **digestive** fluids, scorpions melt their prey. The scorpions then drink the food like smoothies.

**Scorpions can go a whole year without food.**

# CHAPTER 3

# FAMILY LIFE

Most scorpions live alone. They only meet to **mate**. Male scorpions search for females using scent.

Two to 18 months after mating, females have babies. Newborns look like mini adults. They climb onto their mothers' backs. They'll be safe from predators.

## Growing Up

Babies stay with their mothers until their first **molt**. Then they go off on their own. Young scorpions molt about five times. They grow with each molt.

Scorpions are in danger after molting. Their bodies are soft. Their new exoskeletons take time to harden.

# Scorpion LIFE CYCLE

Female scorpions have babies.

Scorpions are fully grown after a few months to seven years.

Babies climb on mothers' backs. They stay there for one to seven weeks.

Baby scorpions molt and grow.

# Scorpion Food Chain

This food chain shows what eats scorpions. It also shows what scorpions eat.

**OWLS** **REPTILES** **MAMMALS**

**SCORPIONS**

**INSECTS** **SPIDERS** **SMALLER SCORPIONS**

# CHAPTER 4

# THEIR ROLES
## in the World

Scorpions play an important role in the food chain. Many animals find them tasty. Owls and lizards eat scorpions. Mammals dine on them too.

# Super Scorpions

People might think scorpions are scary. But scorpions are important. They eat many insects. And many animals eat them. Scorpions might be creepy. But they are also amazing.

Scientists have found many uses for scorpion venom. Some use it to make **tumors** glow. If a tumor glows, it's easier for doctors to see.

# BY THE NUMBERS

**2 to 8 years**
life span

## NUMBER OF SCORPION SPECIES

### about 2,000

# 1 to more than 100

NUMBER OF BABIES
A FEMALE SCORPION CAN
HAVE AT ONE TIME

## 25
NUMBER OF KNOWN
SPECIES THAT ARE
DEADLY TO HUMANS

# GLOSSARY

**arachnid** (uh-RAK-nid)—a class of animals including spiders, scorpions, mites, and ticks

**digestive** (dy-JES-tiv)—having the power to cause or help digestion

**exoskeleton** (ek-so-SKE-le-ten)—the hard, protective cover on the outside of an insect's or arachnid's body

**mate** (MAYT)—to join together to produce young

**molt** (MOLT)—to lose a covering of hair, feathers, or skin and replace it with new growth

**paralyze** (PAR-uh-lahyz)—to make someone or something unable to move or feel all or part of their body

**predator** (PRED-uh-tuhr)—an animal that eats other animals

**prey** (PRAY)—an animal hunted or killed for food

**thaw** (THAW)—to stop being frozen

**tumor** (TOO-muhr)—an abnormal growth of tissue

**venom** (VEH-num)—a poison made by animals used to kill or injure

**vibration** (vahy-BREY-shuhn)—a quick motion back and forth

## BOOKS

**Blake, Kevin.** *Deadly Scorpion Sting!* Envenomators. New York: Bearport Publishing, 2019.

**Nugent, Samantha.** *Scorpions.* Fascinating Insects. New York: AV2 by Weigl, 2017.

**Sullivan, Laura L.** *The Deathstalker Scorpion.* Toxic Creatures. New York: Cavendish Square Publishing, 2017.

## WEBSITES

Fun Scorpion Facts for Kids
**www.sciencekids.co.nz/sciencefacts/animals/scorpion.html**

Scorpion
**kids.nationalgeographic.com/animals/scorpion/#scorpion-tail-up.jpg**

Scorpions
**www.ducksters.com/animals/scorpion.php**

# INDEX